Grasshoppers

ABDO
Publishing Company

A Buddy Book
by
Julie Murray

Published by Buddy Books, an imprint of ABDO Publishing Company, 4940 Viking Drive, Suite 622, Edina, Minnesota 55435. Copyright © 2005 by Abdo Consulting Group, Inc. International copyrights reserved in all countries. No part of this book may be reproduced in any form without written permission from the publisher.

Printed in the United States.

Edited by: Christy DeVillier
Contributing Editors: Matt Ray, Michael P. Goecke
Graphic Design: Maria Hosley
Image Research: Deborah Coldiron
Photographs: Corel, Flat Earth, Image Ideas, Ingram Publishing, Mark Kostich

Library of Congress Cataloging-in-Publication Data

Murray, Julie, 1969-
   Grasshoppers/Julie Murray.
      p. cm. — (Animal kingdom. Set II)
   Contents: Insects — Grasshoppers — Kinds of grasshoppers — What do grasshoppers look like? — Where do grasshoppers live? — Noise makers — What do grasshoppers eat? — The beginning of a grasshopper — Nymph to adult.
   ISBN 1-59197-317-1
      1. Grasshoppers—Juvenile literature. [1. Grasshoppers.] I. Title.

QL508.A2 M95 2003
595.7'26—dc21

                                                    2002038561

Contents

Insects

There are more than one million kinds of insects. All insects have six legs and a pair of **antennae**. Antennae are the long feelers on an insect's head. Butterflies, bees, ants, and grasshoppers are all insects.

There are many kinds of insects.

Grasshoppers

Grasshoppers have been around for millions of years. Today, there are more than 18,000 kinds of grasshoppers.

There are many kinds of grasshoppers.

There are two main groups of grasshoppers: short-horned and long-horned. Short-horned grasshoppers have short **antennae**. A common short-horned grasshopper is the locust.

A locust is a short-horned grasshopper.

Long-horned grasshoppers have longer **antennae**. A common long-horned grasshopper is the katydid.

A katydid is a long-horned grasshopper.

What They Look Like

Most grasshoppers are green or brown. Some have markings that are black, red, or yellow.

Some grasshoppers have bright colors.

Some grasshoppers grow to become as long as six inches (15 cm). Others are shorter than one inch (three cm).

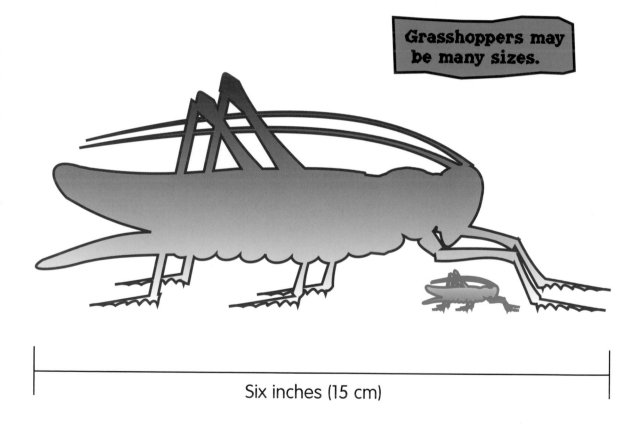

Grasshoppers may be many sizes.

Six inches (15 cm)

Grasshoppers have three main body parts: the head, the **thorax**, and the **abdomen**. On the grasshopper's head are five eyes and two **antennae**. Grasshoppers use their antennae to feel what is around them.

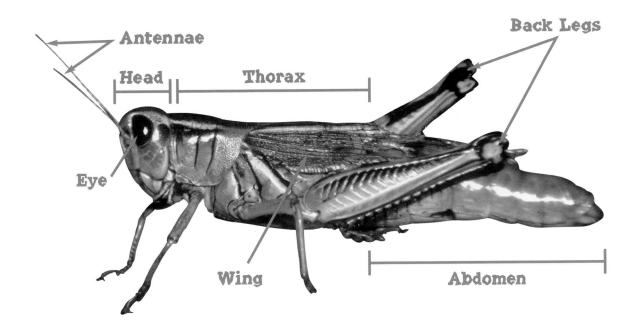

Antennae

Back Legs

Head

Thorax

Eye

Wing

Abdomen

The **thorax** is the middle part of a grasshopper. On the thorax are wings and six legs. The large back legs are for hopping. Grasshoppers can hop very far for their size. Some can jump 20 times the length of their body.

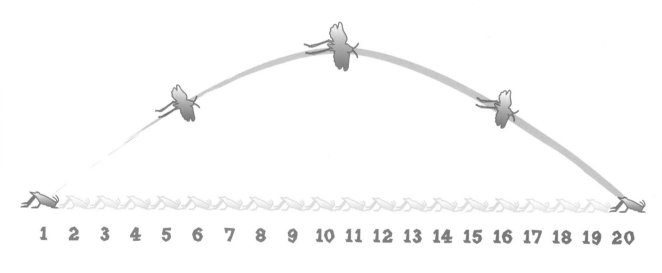

1 2 3 4 5 6 7 8 9 10 11 12 13 14 15 16 17 18 19 20

Grasshoppers can go far in one hop.

Where They Live

Grasshoppers live in many places around the world. They mostly live on grasslands. Grasshoppers also live in mountains, deserts, and gardens. They go where there is food, sunlight, and soil. A field of grass and flowers is a good place to find grasshoppers.

Eating

Grasshoppers mostly eat plants. They eat flowers, vegetables, leaves, and seeds. Grass is a favorite food.

This grasshopper is eating a flower.

Some grasshoppers travel and eat together in large groups. This is called swarming. Swarming locusts can be pests. Sometimes they swarm through farmland and eat all the crops.

Swarming locusts can hurt farm crops.

Staying Safe

Grasshoppers must watch out for spiders, praying mantises, birds, and snakes. These **predators** eat grasshoppers.

A praying mantis

Hiding is one way grasshoppers stay safe. Grasshoppers can hop or fly away from danger, too. Some grasshoppers make a liquid when they are scared. This bad-smelling liquid helps keep **predators** away.

Having wings that look like leaves helps this grasshopper hide from predators.

Singing Grasshoppers

Grasshoppers make sounds, or "sing." Some grasshoppers sing by rubbing their front wings together. Other grasshoppers sing by rubbing a back leg against a front wing.

Most singing grasshoppers are males. They often sing to find mates. Sometimes, male grasshoppers sing to other males. They are warning them to stay away.

Front Wing

Back Leg

Grasshoppers use their legs and wings to make "singing" sounds.

Each kind of grasshopper has its own song. Many short-horned grasshoppers sing during the day. Katydids mostly sing at night. Late summer and early fall are the best times to hear grasshoppers.

Stages Of Life

A female grasshopper lays her eggs during the fall. She lays them in a hole in soil. A grasshopper may lay as many as 100 eggs at one time. She covers the eggs with thick foam. This foam hardens into an egg pod.

Grasshopper eggs commonly hatch in the spring. The baby grasshoppers have no wings. They are called **nymphs**.

A grasshopper nymph

The **nymphs** hop around and eat. They grow very fast. The nymphs shed their skin, or **molt**, many times. A nymph is full-grown after molting five times. As adults, grasshoppers may live for three months.

Nymphs grow wings and become adult grasshoppers.

Important Words

abdomen the back end of an insect.

antennae the two long, thin "feelers" on an insect's head.

molt to shed and grow new skin.

nymph a young grasshopper.

predator an animal that hunts and eats other animals.

thorax the main middle part of an insect.

Web Sites

To learn more about grasshoppers, visit ABDO Publishing Company on the World Wide Web. Web sites about grasshoppers are featured on our Book Links page. These links are routinely monitored and updated to provide the most current information available.

www.abdopub.com

Index